OLD WIVES' AND DOGS' TALES

OLD WIVES' AND DOGS' TALES

The Natural Way to Pet Health
Using Traditional Country Cures

by
Linda Adam

BROADCAST BOOKS

First Published By Broadcast Books 1998, reprinted 2000

4 Cotham Vale, Bristol BS6 6HR

Printed and Bound in the UK by Alden Press,

Osney Mead Oxford

Cover Design by Sava Fratantonio

DEDICATION

I would like to dedicate this book to Wally Dewsnip and John Hellings. Without their encouragement and help this book would never have got off the ground. Also to my husband Michael, who worked so hard and for so long trying to sort out my scribblings and to put them into understandable English on the computer.

DISCLAIMER

The purpose of this book is to help with minor ailments and the day to day mishaps that do not require the immediate attention of a vet. The contents of this book do not in any way replace veterinary advice. If you are in any doubt about the health of your dog you must take it to a vet.

CONTENTS

AN ALPHABETICAL LISTING OF AILMENTS AND THEIR CURES

HOW THIS BOOK CAME ABOUT

Like most things in life, this book came into being by accident.

I had always been interested in alternative ways of treatment in animals, as there have always been animals in my family. When I started going to horse shows, and then to the local goat club, I was fascinated to hear the stories that the older and much wiser stockmen and women had to tell about how they had treated animals over the years - especially during the war when money and sophisticated medicines were scarce.

I thought then that here was a body of folk knowledge that had been handed down for generations, and that should not be forgotten in an age when over the counter 'cure-alls' are only too readily available.

I started to attend dog shows and discovered many fascinating tips in the same way. I soon found myself writing everything down, and trying the cures out on my own dogs with, may I say, the greatest success. Too many times, these cures were really helping my dogs' symptoms for the 'Old Wives' Tales' to be dismissed as old rubbish.

✳✳✳✳✳✳✳✳✳✳✳✳✳✳✳✳✳✳✳

For the last few years, I have been contributing these tried and tested tips to a publication called *'South West Dogs'*, and readers in turn sent me their own tips. Listeners to my subsequent radio spot on Severn Sound Radio did the same, and my collection began to grow. Whenever I gave a talk at seminars, outlining a health problem in dogs, someone would invariably come up to me later with a remedy. The collection grew rapidly, and what you are about to read is the result of many requests I have received to put these tips together in one book.

I owe a great thank you to all the people who have so generously shared their knowledge and experience with me over the years. I hope you will find this collection as helpful as I have done - and if you have any further tips to add, please let me know!

Linda Adam 1998

Jasmine Cottage, 103 Westbury Leigh

Westbury, Wiltshire

✳✳✳✳✳✳✳✳✳✳✳✳✳✳✳✳✳✳✳

****

A NOTE ON INGREDIENTS

Do you know that dog owners spend over forty million pounds in this country last year? This huge outlay is unnecessary. We can save a lot of money by just using what is in our household cupboards or what is growing in our gardens and fields. In the following pages, certain simple ingredients will crop up again and again.

One of my favourites is **bicarbonate of soda**. A good pinch of bicarbonate of soda in your dog's food will reduce the smell of your dog's urine, for example. If you read on you will see how this wonderful substance can also be used for teeth cleaning, as an antidote in poisoning, for keeping white dogs bright, as a cure for heat-stroke, coat chewing, paws licking, removing smells, and in many other ways.

Cider vinegar is a wonderful disinfectant and is effective against fleas, carpet smells. It can be used to help with rheumatism and as a coat conditioner for dogs. **Cornflour** helps with burns, **flour** with bleeding nails. You will also find many uses for **salt, garlic, lemons**, and **pepper** in this book.

The herbs I have mentioned are easily available or grown. The following herbs have proved most useful to me:

****

Parsley cut up finely (you can do this effectively with scissors in a jar) and sprinkled on your dog's food, will aid with cystitis and prostate problems and will also aid in keeping your dog's breath fresh. It is also used for carsickness in dogs.

Dandelions have many uses, and can be infused to make a tea. This makes a good tonic (as does nettle tea), helps with jaundice and colic, and cleanses the body of toxins. Just put a couple of drops in the dog's food. Remember the 'milk' from the stems is also very good for clearing warts. Apply the sap once a day until the wart disappears. If you haven't any dandelions in your garden you can find them on waste ground, or can even grow then in a pot, by sowing the fluffy seed heads in compost. Gardeners will think you have gone mad...

Everyone knows how much cats love rolling in **catnip,** but it is also excellent for clearing fleas, as is **fleabane, mint, tansey** and **lemongrass. Chickweed**, which is found on the drier edges and areas of grassy wasteland, has many uses. To make chickweed tea, take 1oz of plant, and infuse with 20 fluid ounces of boiling water poured over the chickweed. Cover and leave to cool, then strain, bottle, and store. Use this infusion for treating eye problems, rashes, stings, and skin allergies. **Comfrey** solution mixed with soluble aspirin given once a day will help with arthritic pains. Many of us have

marigolds in our gardens. If not, they are very easy to grow. The flower of this plant is very useful for blisters, eczema, and stings, not only for dogs but also for us). Steep 1oz of fresh marigold petals in 1 pint of boiling water, cover and leave until cold. Strain the liquid and place in a container. Use this to dab onto blisters, eczema, or stings for instant relief. This mixture will keep in the fridge for up to 3 months. You could also freeze it to thaw out at a later date when needed. **Thyme** will help with dandruff. These are the herbal uses you will read about in this book; a herbalist will have hundreds more.

The wood engravings of **Bosie,** and of the fruits
and flowers in this book, are by Hilary Paynter.

OLD WIVES' AND DOGS' TALES

An Alphabetical Listing of Ailments and their Cures

ACHES

Rheumatism, Spondylitis and Hip Dysplasia

You can tell if your dog is suffering this type of inflammatory ailment if it seems stiff after walks or other exercise. You should initially consult your vet.

In cold damp weather, rheumatism in some of our older dogs could be triggered off. The following recipe will greatly help with rheumatism or even a severe sprain. Put one tablespoon (equal to 1/2 fluid ounce or 15 millilitres) of Epsom Salts in boiling water and allow it cool right down. Stand your dog with its leg in the basin and trickle the water over the affected area. Repeat the treatment two or three times a day until you notice an improvement.

If you have an older dog that suffers from any of the above, you can offer a lot of relief by putting 1 teaspoon (1/6 fluid ounces or 1/3 tablespoon or 5 millilitres) of cider vinegar into 1 pint of drinking water daily. This is very helpful for these conditions.

Here is a tonic which can be easily made up and given to an arthritic dog to ease the pain:

Half a grapefruit
1 orange
1 lemon
2 sticks of celery
4 cups of water
1 tablespoon of vinegar
1 tablespoon of Epsom Salts

Cut up the fruit and celery, including the peel, simmer in water uncovered for 1 hour. Press the pulp through a sieve, stir in 1 tablespoon of vinegar and 1 tablespoon of Epsom Salts. Pour one quarter of a cup of this mixture at a time into the dog's normal drinking water. Expect to see an improvement after about 4 weeks of regular use.

This drink will also help with rheumatism: Boil the leaves and any left over bits of fresh celery. When cool, give the water to your dog to drink. If you can give him this water for a couple of times a week over a longer period this will help him greatly.

Also, do not throw away the water in which you have boiled your potatoes. Dabbing this liquid onto the affected joint can bring a lot of relief.

Further relief for aches can be obtained by making a home-made salt bag. Get a very strong piece of material and sew up an open-ended bag. Fill this with salt and then sew the opening up. Place into a microwave or oven until warm, then hold it over the affected site of the pain for as long as you can. The salt will retain the heat for a long time. This is also very useful for a safe alternative to a hot water bottle, as there is no boiling water to leak out.

ACNE

Acne can be wet or dry, and shows up in small patches anywhere on the dog.

Mix half a cup of Milton's sterilising fluid to half a pint of cooled boiled water (no stronger). Apply over the affected areas and allow to dry naturally.

ANAL GLANDS: SWOLLEN

If your dog is regularly wormed (as it should be, every six months) you will be able to tell if anal glands need attention. Dogs will try to slide along on their backsides, and repeatedly turn around while trying to bite their backs and tails.

Pick the whole plant including the root of a buttercup, dry as you would mint, chop up finely and store in an airtight jar. When you need to use put a dessertspoonful in a container, pour on a little boiling water, and allow to cool. Use this as a cooling lotion for anal glands.

Take a double handful of sage leaves, cover with water and bring to the boil and then cool, strain and use the liquid for bathing infected anal glands. You can also dab undiluted cider vinegar on to the affected area.

A good groomer will be able to empty anal glands much more cheaply than a vet will.

ANTISEPTICS

Neat rice vinegar is a very powerful disinfectant - it kills on contact all dangerous bacteria.

Antiseptic Creams

Melt beeswax and oil together in equal quantities. Add a good double handful of fresh chickweed, then pour into tubs, and chill overnight. This is a very effective antiseptic cream for all types of skin problems.

For cuts and injuries, this recipe makes a wonderful natural healing cream:

Heat gently some liquidized or mashed up garlic,

and mix in enough seaweed powder to make a soft paste. Cool completely, and then use a good smear between gauze and hold in place with a bandage. The dressings should be changed every eight hours.

BALD SPOTS

These can develop on an overweight dog that is lying down for long periods.

Take soot (if available) and lard in equal quantities, mix and use it as a cream.

If soot is not available, massage corn oil massaged into the bald spots and add one teaspoonful to the feed once a day.

Rubbing coconut oil into the affected area is also beneficial.

Feeding tinned pilchards two or three times per week along with the normal diet is a useful aid.

Dry out Scots Pine needles and store in a dark place in a paper bag. Place a handful at a time in a cloth bag or a leg of old tights, and add these to the bath water. This helps with hair growth.

Sometimes after a cut or a wound the hair that grows back on the dog grows back a lighter shade than normal. Although this is a purely cosmetic solution, you can dye the colour back by repeatedly and regularly dabbing very strong black tea onto the area with cotton wool. For best results, use the tea whilst lukewarm and leave on the fur to dry naturally.

BANDAGES

There are times when dogs are required to wear bandages, which they do not like. To prevent them removing the bandage, smear with soap.

BARKING

To stop a dog barking whenever someone arrives at the door, get an empty soft drinks can and place a few coins inside, then tape up the gap in the top. All you have to do when someone knocks at your door is to shake the can vigorously and say in a commanding voice "no". After a few times he will soon get the message! You could occasionally reinforce your dog's 'good' behaviour after shaking the can with a small treat.

BATHING

The easiest way of staying dry whilst bathing the dog is to get a black bin liner with a hole in the bottom to put over your head, and a holes in each side for your arms. No more wet clothes!

Bathing: Soap Free Soap

If you need to wash your dog's face and you are afraid of using soap or shampoo, you can make up your own tearless substitute, which will clean just as effectively. Get some very fine oatmeal, tie it into a cotton square - a handkerchief will be fine, dunk it into water, squeeze out the excess water and use it as you would a normal washcloth. This is a very good "soap" to use on dry skin as well.

BEREAVEMENT

Unfortunately if we keep dogs there comes a time when we have to face up to the fact that our dog is going to die. This can be devastating for some people, especially for the older person who has possibly outlived the family, and all that remains is the dog. You can't always find a friend with a sympathetic ear to listen to you or even understand what you are feeling, or maybe you feel that if you did speak to someone they would think that you were silly in some way. Get in touch with the *Pet Loss Befriendment Service*. Details are on p. 94.

BLACK-FACED DOGS

To freshen and clean black-faced dogs, or to

smarten them before a show, rub with coconut oil soaked in cotton wool.

BLOAT

Bloat is usually caused by a dog eating too fast; causing the stomach to swell with gasses. It is painful and can be serious. Encourage your dog to eat more slowly by giving frequent, smaller portions of food. A cure for bloat is to feed a small portion of live yoghurt immediately after each meal, which will keep the gasses down that cause the harm.

BOILS AND CYSTS

A good natural poultice for cysts is to heat up some milk, into which some breadcrumbs have been crumbled. Apply onto the affected spot for 5 minutes, twice a day until the cyst bursts.

When dogs suffer from boils and cysts we cannot use conventional compresses in case they lick them and become ill. If your dog will keep a dressing on you can use any of the following: a heated slice of tomato, a slice of raw onion, mashed garlic, the outer leaf of a cabbage or an old used tea bag.

Whichever one you choose, make sure that you change it every four hours.

Tea tree oil from health food shops sprayed in between the toes daily will effect a cure within a week.

BRUISES AND SPRAINS

Rub a fresh witch hazel leaf onto the affected area to relieve pain.

Pick the whole plant, including the root, of a buttercup, use two plants and cook in a little water. Cool and discard the plants, then use the remaining liquid as a cooling lotion for bruises.

Relieve pain by using vinegar dabbed onto the bruise with cotton wool.

BURNS

Minor burns and sunburn can be eased by any of the following natural remedies:

Add enough water to cornflour to make a paste and apply directly to the burn, leave until it falls off naturally.

Raw cucumber, potato, or apple should be held in place for 10 minutes.

Yoghurt can be patted gently onto the burn.

A cold tea bag or cold tealeaves can be held onto the burn for a few minutes.

━━━━━━━━━━━━━━━━━━━━━━━━━━━━
✳✳✳✳✳✳✳✳✳✳✳✳✳✳✳✳✳

You could also make up a lettuce leaf mixture and leave it in your fridge for the summer months. Boil several lettuce leaves in a couple of pints of water, strain, then cool for several hours in the fridge. Dip cotton wool into the liquid and gently press onto the affected area.

If your dog has contracted sunburn, rub a fresh strawberry over the affected area to give relief. A fresh raspberry may also be used. Mashed cooked cold or tinned rhubarb rubbed over the affected area will also bring relief from sunburn.

CARS: HATCHBACK

A lot of us carry dogs in the back of cars but it can be hazardous. When a friend of mine got her dog out of the car recently, the dog's paw was cut on some sharp metal inside the car. Check the back interior of your car, and if you find any sharp exposed bits just cut out a length of rubber hose to fit, then slit the hose along its length and this will neatly cover the offending strip of metal safely. It will protect your dog and your hands.

CAR SICKNESS

There are several ways you can combat this. First, try a small chain attached underneath the bumper so

✳✳✳✳✳✳✳✳✳✳✳✳✳✳✳✳✳

that it touches the road. The sound of the trailing chain seems to alleviate car sickness.

You can also sit your dog on newspaper. This sometimes helps, or if not, will at least deal with the mess! Most young dogs usually grow out of carsickness but if yours does not, here are a few things that you could try:

Take half an inch of fresh ginger, peel it and grate it into a teacup, pour on boiling water, cover and leave to infuse for 5 minutes. Strain, then give your dog a quarter of this mixture 1 hour before travelling.

This is very good for human carsickness and morning sickness as well, but in that case, an adult would drink the whole cupful.

If your dog won't drink the ginger mixture, you could try a small ball of cornflour moistened with water. Feed this half an hour before travelling. It is very palatable and absorbs excess saliva. Additionally, put a bunch of fresh parsley in your dog's travelling box. The aroma can also help.

Crush a fresh angelica leaf and put near the airflow of a car.

CHEWING COAT AND LICKING PAWS

Some dogs develop an unpleasant habit of chewing

their coat, usually on the paws.

First, make sure that this is just a habit and is not done for any other reason, such as an infection or a foreign body in the paw. Another cause may be a slight allergy: some cotton or nylon loop carpets can burn the soles of dogs. If you are sure there is no physical reason for the chewing, try this cure:

Take 1 tablespoon of bicarbonate of soda to 1 pint of boiling water, and allow to cool for a couple of minutes. Then soak your dog's feet in this for about a minute, and allow to dry naturally.

Another cure is to put some washing up liquid on the paw and liberally sprinkle hot cayenne pepper on it. Once your dog tries to chew this concoction, it will soon stop.

Some dog owners with dogs such as West Highland Whites, Bichons, Poodles etc., get problems with their dogs licking their feet, which turn pink. Although this is harmless, it looks unsightly. Try a teaspoonful of bicarbonate of soda in the dog's drinking water. This will help to neutralise the acidity of the saliva, which is the cause of the paws going pink.

CHEWING FURNISHINGS

This is an age-old problem. There are lots of products on the market to help prevent this, but I find the best thing to use is a green chilli, cut in half and rubbed on the affected piece of furniture. (Do not use a red chilli as this can cause digestive upsets). A green chilli is fine and won't leave a stain on the furniture. The best thing of course is not to leave a dog or puppy alone long enough for it to become bored and want to start chewing.

Some of you may have a new puppy and perhaps it is beginning to "attack" the furniture, leaving teeth marks in wood. To stop this before there is any further damage, you will need something that will not stain the wood or harm your puppy. Go to your chemist and buy some Oil of Cloves. (It is very cheap, and excellent for numbing a gum sore with toothache). With a tissue, wipe some oil on the wood that the puppy tries to chew. It will dislike the smell and the bitter taste.

Another substance you could try if you want to stop your dog from chewing your woodwork or your furnishings, is to smear the surfaces with a hot peppered sauce like tabasco. This is not harmful to your dog; and usually the smell of the tabasco is discouragement enough. It will also easily wipe off the surfaces. Vick is another good deterrent. Even better is to encourage your dog to chew on its own playthings by smearing toys and bones with peanut butter, which dogs love.

CHEWING GUM

In hot weather, our dogs sometimes bring in chewing gum on themselves, which gets transferred to bedding or onto furniture.

An easy way of removing this, if the gum is attached to a small article, is to place it in your freezer until it is frozen and then chip it off. If the article is too big to go into the freezer, hold anything frozen onto it until it is chilled. Saturate the gum with warm vinegar, leave for 5 minutes, and then brush out.

CHEWS FOR DOGS (HOME MADE)

It is very hard to find suitable chews for our dogs' amusement. Either they eat them too quickly or they splinter and become dangerous. You can make your own dog chews that are very safe, and dogs love them.

At last there is a use for tough old rose bush roots! Cut off the tops down to the rootstalk, prune off any long or thin roots, and then scrub the root well. Leave it to dry and harden for two or three weeks then give it to your dog to play with.

You could spread bacon fat, dripping or peanut butter on it for extra taste. This chew does not splinter, and dogs love them.

✳✳✳✳✳✳✳✳✳✳✳✳✳✳✳✳✳✳

CHILLS

If your dog is trembling, and his ears feel cold to the touch, he has probably caught a chill. If your dog lives outside in a kennel, or if you have both just come back from a cold walk, give a small drink of evaporated milk and warm water with a spoonful of honey. This will keep chills at bay.

We can also help our dogs by making them a home made health-giving drink out of pearl barley. This is good for the skin, helps purify the blood and is especially good for kidney and bladder conditions:

Bring 1 pint of water to the boil and add one table-spoonful of pearl barley, simmer for 20 minutes, cool, and strain and add one teaspoonful of honey.

A spot of Eucalyptus oil under the dog's chin will make breathing easier. Once or twice a day should be enough.

COLIC

Colic is a severe, spasmodic, griping pain in the belly. To relieve the symptoms, simmer 2ozs clean fresh dandelion roots in water, strain through a sieve and store in the fridge. Give in doses of one tablespoon three times a day. This will keep for three days.

COLITIS

This is an inflammation of the lining of the stomach, caused by intolerance to certain foods. Diarrhoea is one of the symptoms. It is a condition that must be confirmed by a vet.

To help the symptoms, mix one pint of litterlac (available from chemists) with two tablespoons of arrowroot.. Heat until boiling whilst stirring well, and cook for one minute. Remove from the heat and allow to stand for five minutes, then beat in two raw eggs.

Feed this three times a day for three days, then half the mixture added to cooked chicken and boiled rice for two days. Try normal food with half a pint of water and one tablespoon of cooked arrowroot for three days. When the stools are normal, stop the arrowroot.

Another natural cure for colitis is to take a quantity of stinging nettles or any strong green vegetable and put then in a pan, just covered in water.

Bring to the boil and simmer until cooked, drain, add two or three slices of brown bread, and liquidize. Add about a tablespoon of brandy, and give the dog as much as it will eat, every two hours, until there is an improvement in the stools.

CONSTIPATION

If your dog has not passed a motion in 48 hours, your vet MUST be consulted.

You will find any of the following useful:

Half a chopped fig, fresh or dried, added to your dog's diet daily.

A few walnuts crushed to a fine powder and mixed in with the dog's food, one teaspoon at a time.

Tinned apricots, a couple of slices at a time.

A dessertspoonful of cooked or tinned rhubarb added to the food.

A teaspoon of linseed oil in the diet.

Small pieces of children's liquorice.

Plain live yoghurt twice a day, a dessertspoonful at a time.

CUTS

Never mind how well we look after our dogs, they will get a minor cut from time to time. I keep a small jar of crushed eggshells in my kitchen cupboard just for this kind of accident. Wash and dry a few eggshells, then crush them to powder using a liquidiser or a mortar and pestle. Store the powder in an airtight container where they will keep indefinitely. When a small wound appears on your dog, sprinkle some powder into the wound and the blood will clot immediately.

Turmeric powder will also stop the bleeding of cuts.

A spider's web placed over the wound will soon stop the bleeding.

(For discoloured hair growing back over a cut, see **Bald Spots**)

CYSTITIS

This is an inflammation of the urinary tract, which is painful and unfortunately crops up all too often. Signs of cystitis in your dog are showing pain on urination, and frequent urination of small amounts of urine. Even if your dog is under medication for this, you can use the following with no side effects: obtain some dried hops, then powder them as finely as you can, store in an airtight jar and feed one teaspoonful at a time in food per day until the symptoms clear. To help prevent this recurring, try a small pinch of salt on the food each day.

Fresh parsley, cut up finely and sprinkled onto your dog's food will aid with cystitis and prostate problems and will aid in keeping your dogs breath fresh.

Although asparagus is expensive, this cure uses the

tough, lower part of the stem, which is usually discarded. Cook and finely chop, then add to the diet one dessertspoonful at a time daily for about a week.

If you live by the sea or visit the coast, pick some wild sea holly, dry as you would mint, then crush to powder and store in an airtight jar. Use a teaspoon at a time in your dog's diet to aid cystitis and urine infections.

Cover 4 ozs of pearl barley in a little water and bring to the boil, strain and throw away the water. Pour 1 pint of boiling water over the barley and simmer for 5 minutes, add the zest of a lemon and allow to cool. Give this liquid to your dog to drink as often as necessary.

DANDRUFF

Small white specks in the coat are signs of dandruff. The following is a very effective natural spray:

Take 4 heaped tablespoonfuls of dried thyme and boil up in a pint of water for 10 minutes. Strain the mixture and allow to cool. Put into a spray bottle and spray lightly onto your dog's coat, massage in and leave to dry naturally.

Use this treatment twice a week until the dandruff goes.

DIARRHOEA

If your dog has developed diarrhoea, what you need to do is to starve your dog for twenty-four hours, giving only mineral water to drink and a teaspoonful of live yoghurt three times a day. This will help to readjust the natural bacterial balance in the intestine. However, if the diarrhoea persists for more than forty- eight hours or worsens, seek the advice of your vet.

Cook and mash a few sweet chestnuts and add a heaped teaspoonful to your dog's diet daily to aid recovery. One teaspoon of lime juice in the dog's diet will also help.

DIABETES MELLITUS

Diabetes is a disorder of the pituitary gland making the body unable to absorb sugar effectively. The signs are excessive urination and thirst. Your vet must be consulted if you suspect diabetes. The following measure will also help: feed entire dandelion plants to your dog, to help it retain insulin.

DIETS

Coprophagia (Eating Faeces)

Not a very nice subject to talk about! Put a slice of

pineapple or three pineapple chunks in the dog's food daily, or 1 tablespoonful of pure natural pineapple juice. For some reason, the dog hates the smell of digested pineapple in its droppings. You will need to do this for at least a fortnight before you see any signs of this working.

You could also try adding 1/2 a teaspoonful of cider vinegar to the diet each day. These are independent treatments to be tried separately, not at the same time. If the above two do not work, try a cherry-sized piece of brewers yeast daily.

A piece of kipper in each meal or a sprinkling nutmeg over the dog's food has also been known to work.

Courgettes can also be used. Cut them into small pieces, cook them in a little olive oil, and add one dessertspoonful to the diet each day.

Lack of Appetite after Illness

Try giving your dog chicken-flavoured COMPLAN to begin with, and then try chicken korma (believe it or not, dogs love it!)

Loss of Appetite

Add 2 or 3 drops of wild Marjoram oil to drinking water daily.

You could also try giving double the recommended

dose of brewers' yeast for ten days.

(See also: **Thin Dogs**)

DIGGING

One solution which has been quite successful is to blow up a balloon and then dig a hole where the dog digs regularly, wide enough and deep enough to bury the balloon and cover it with soil. When the dog digs his hole again the balloon will burst and, with luck, the bang of the balloon bursting will put your dog off digging again, without harm.

If you are trying to protect young seedlings, cover the spot with a piece of weld mesh or similar. Failing that, you could always cover the area to be protected with holly or any other thorny cuttings.

DRINKING WATER

If, like me, you collect rainwater for your dogs to drink, you will want to make sure that it is safe for them. A simple way to ensure that it is pure is to pour some water into a glass last thing at night, and place a sugar lump into it. Leave the glass in a warm room overnight. If the water is clear the following morning, the water will be quite safe, but if the water is cloudy or milky do not risk your pets drinking it.

DROOLING

Some breeds, especially those with pendulous cheeks, naturally drool more than others do.

Put 1 drop (only) of oil of cloves onto a sugar cube, and feed to the dog just before going out.

Another tip is to give the dog a small piece of crystallised ginger.

A tablespoon of camomile, lemon balm or peppermint tea can also help.

EAR MITES

If your dog is scratching its ears a lot, suspect mites. Dogs are especially prone to them in spring. You can make up your own very safe ear oil to help with this problem. Crush four cloves of garlic and let them steep in one cup of olive oil overnight, then discard the garlic. Heat the oil until it is just barely warm and put several drops into your dog's ear. Do this every other day until the condition improves.

EAR SCRATCHING

If the ear scratching is only due to dirty or slightly waxy ears, you can gently clean them with cotton wool soaked in alcohol to dissolve the wax and dirt. A drop of mineral oil on cotton wool will also clean them, and calamine lotion will refresh the ear.

ECZEMA

If your dog has eczema or itchy skin, one way of successfully treating this is to give your dog an oat bath. Get a muslin bag or one leg of an old pair of tights, and pour roughly 1lb of porridge oats into this, making sure the bag is securely tied at both ends. Soak the bag in a bucketful of tepid water and rinse your dog thoroughly with this solution, making sure you soak the fur right to the skin. Then let your dog dry naturally.

Also, try equal amounts of parsley and watercress in the food once a day or two Milk of Magnesia tablets morning and night for three days.

Surgical spirit is also very helpful. Dogs sometimes suffer with red patches between their toes or pads: surgical spirit sprayed on daily will usually clear it up.

Applying a paste made from cornflour and vinegar can ease itchy swellings. Pat this onto the affected area and the irritation will be drawn out.

Rub a mallow flower or leaf over the affected area regularly until a difference is noticed.

(See also: **Skin**)

ELIZABETHAN COLLARS

When a dog has been injured, an Elizabethan collar is sometimes necessary to stop it getting at the stitches or dressings. You can save the expense of buying one and make your own.

Cut a large square of cardboard from an old box, then cut a hole in the centre. Make a slit up to the centre in the middle of one side. Slip this onto the dog's head and tape up the slit. You can make as many of these as you need for next to no cost. I have also seen a dog peeping out of the rim of a plastic bucket with its base removed.

EYES

If your dog has sore or red rimmed eyes, a really good soothing recipe is a teaspoon of olive oil to a teacup of cold tea. Mix well to 'infuse'. Using cotton wool, bathe the affected part. This will make the skin more supple and encourage hair grow.

A cold, used tea bag squeezed gently over the eyes also works well.

If your dog is out on a breezy, very dry day, dust and grit can be blown into the eyes. If this happens to your dog and you are absolutely sure that there is nothing else wrong with its red, sore eyes, you can make your own safe eye wash. Extract the juice from a small piece of fresh cucumber and drop into each eye three times a day until the redness disappears. The cucumber must be fresh for each occasion.

Weak Eyes

Some dogs are prone to watery weak eyes. If this is a problem, use pure cod-liver oil as eye drops daily. It will not cure the problem but it will keep it under control.

FEET

Burnt Pads

Be very careful where you let your dog walk in the hot weather. When tarmac surfaces melt, the tarmac will get into your dog's paws, and could burn and peel the skin. If you notice that your dog has picked up some melting tar on its pads, get some cooking oil - and gently rub the tar away using cotton wool dipped into the cooking oil. The tar will dissolve. Then soak the paws in salt or vinegar dissolved in water. Smear the pads with a good antiseptic cream, and all should be well. Use cooking oil to remove oil or tar from your dog's fur as well.

Soft And Cracked Pads

If your dog's pads are particularly soft and prone to

cracking, wait until they have healed and every so often (on a healthy paw) dab on surgical spirit with a wad of cotton wool. This will harden them up. Doing this about three times a week should do the trick.

For sore cracked pads, boil up some potato peel and use the water when cool to immerse the paw. The pad will heal in a couple of days.

Trim out as much of the hair between the pads as possible and dust with an anti-fungal powder.

Smear vaseline on wet pads on your dog, then dust liberally with baby powder to create a covering on the pad.

Neatsfoot Oil (from agricultural suppliers or horse tack shops) rubbed into the paws will also soothe sore pads.

You can also use 1 pint of ivy leaves boiled in 2 pints of water. Allowed to cool, and add 1 dessert-spoonful of Witch Hazel to every cupful of liquid. Bathe this onto the paws, then dust them with very fine oatmeal.

It may help to stand your dog in warm water with two or three tablespoons of Epsom Salts added for about 15 minutes twice a month.

If your dog gets a wood splinter in its paw, soak the

area in cooking oil to soften the skin, and then apply an ice cube to deaden the paw before removing the splinter with tweezers.

Salt on Feet

After the winter there is often a lot of salt left on the roads, especially if it has been dry. If dogs get this in their paws, it can become very irritating. If you suspect that your dog has got salt in its feet, rinse the paws with warm water (no soap) then soak each paw in a vinegar solution of 1 part vinegar to parts water for about 1 minute per foot. Let the feet dry naturally.

FERTILITY IN DOGS

To help increase the fertility of dogs and bitches, feed a teaspoon of hemp seed oil (available from agricultural suppliers or horse tack shops) in their food every day for at least a month before mating.

FINDING TINY TABLETS AND PINS

Sometimes the tablets of an animal's medication are very small, and once dropped on the floor are impossible to find. The same goes for needles and pins, which our pets could get in their paws. The solution is to fit a pair of old tights onto your vacuum cleaner hosepipe with an elastic band. Vacuum the floor and keep checking the end of the hose. Anything small will become trapped on the end and then you can be sure that you have picked the tablet or pin up.

FLATULENCE

This is a very common problem. Try to make your dog eat more slowly by placing a large pebble - much to large to swallow - in the feed bowl to encourage it to eat more slowly and to swallow less air. Raising the feeding bowls off the floor will also help, because if the dog is not bending down so far to eat, it is swallowing less air.

One other thing that can be tried is to obtain some live goats milk yoghurt from a health food shop, mix a dessertspoonful with your dog's food. This should help considerably.

A teaspoonful of pure orange juice on its own will help with wind and colic, given as and when necessary.

A freshly grated radish sprinkled over the food will also help with wind.

FLEAS AND LICE

Fleas are the culprits if you see tiny black spots on

your dog's coat. This is the dried blood from flea-bites. Your dog will also be scratching often, although this may not always be the case. Some dogs are driven mad by fleas; others hardly seem to notice them.

Sprays

A very effective way of ridding your dog of fleas is to rinse it off with a vinegar rinse (1 part vinegar to 2 parts water) after bathing. This will kill fleas and leave no smell.

For day to day protection, mix up a cider vinegar solution of 1 part vinegar to 2 parts water, keep it in a spray bottle (the little hand held ones, the type that you can buy in garden centres) and spray you dog's coat daily. Vinegar - at around £1 a gallon - must be the cheapest and safest and one of the best insecticides we can buy.

Boil a good double handful of elder leaves in water, cool, and strain; put the water into a spray bottle. This solution is a very good natural flea or insect spray.

Lice

These tiny creatures look like dandruff, in fact they are sometimes called "walking dandruff". You will have to look very hard to see them: they usually cling to the ears.

Lice are even worse than fleas; the only advantage with them is that they stay on the host animal. To discover whether your dog has mites, look very carefully around the edges of your dogs ears, part the hair and examine the skin, using a magnifying glass if you have one. If you see any, dab cotton wool soaked with neat vinegar onto the skin, or dip a flea comb in vinegar and comb through the coat. Do this every day until they disappear. You can use ordinary malt vinegar, but if you don't like the smell use cider vinegar instead. Incidentally, did you know that the vinegar rinse used on our own hair would make it shine?

A natural way of keeping our dogs bedding free of fleas is to pick some mint. Make up little sachets just like the old fashioned lavender bags and place them under your dogs bedding. This will deter fleas from living there.

If you use flea collars on your cat or dog, you will always have a length left over when you cut them to size. Don't throw these bits away; put them into your vacuum bag and this will kill any insect that is sucked into the vacuum cleaner.

If you can give your dogs garlic once a day it will definitely help with preventing worms and fleas. You can buy garlic tablets in chemists, but you can also make your own preparations:

Flea Spray

Take 1oz of fresh crushed garlic to 20 ml of boiled water, put in a glass jar after cooling and seal. Add 2 drops to your dog's food each day. Made this way it will keep in the fridge for about two months, or alternatively you can freeze it into ice cubes and use as required. You can use this recipe as a very effective flea spray.

Brewers yeast powder rubbed straight into the dog's coat, massaged and left, also helps against fleas.

Lavender Essential Oil rubbed into the coat will also help. Treat your dog internally as well, by adding fresh garlic to your dog's food daily. Just 1/2 a clove will do. This will take about 3 months to work.

A Home-made Flea Collar

Soak a large handkerchief in the following mixture overnight: one drop of oil of lavender, cedarwood, thyme and fennel, half a teaspoon of alcohol (vodka is very good) all mixed with the contents of four garlic capsules and one drop of garlic essential oil. Dry and tie around your dog's neck with a knot that cannot slip. This natural scarf will repel fleas for about two months. There is no need to throw it away - just repeat the process. If all these ingredients deter you, soaking the handkerchief in oil of lavender only will work as well.

Another good flea deterrent is to rub the citrus oil from orange peel, or essential rosemary oil, onto your dog's coat.

A home-made flea collar can also be made from some strong fabric stuffed with the herbs tansy and catnip.

To kill lice eggs (nits) wash your dog once every two weeks in a five per cent solution of white vinegar and water.

Flea Traps

Get a night-light or a bedside light with a low wattage bulb and plug it in on the floor in a corner of the room that you want to treat. Under the light, place a wide container about 4 ins deep, with 3 ins of water in it. Fleas are attracted by warmth, and whilst you are sleeping overnight the fleas are quite happily drowning themselves! This flea trap will work just as well as flea shampoo, and costs you next to nothing.

If you still are getting bothered by fleas in the home (and it does seem to be an all year round problem now), buy an old fashioned fly paper and cut into strips and place in the corners of your room. This is very good for trapping fleas, and is completely safe.

We can use natural and safe flea killers in our homes as well.

Herbs such as pennyroyal, tansy, and fleabane are excellent for sprinkling around the home. Use them fresh or dried sprinkled around the edges of your rooms - also in your dogs bedding - leave for a few hours, and them vacuum up. You can leave the herbs in your dog's bedding for about a month to remain effective.

You can also make your own fumigator. Get a small, heat-proof container and cut up some lemon grass. Close all the windows and light the lemon grass, leave it until it goes out then vacuum your room. Repeat as often as necessary.

Fleas in carpets

To discourage fleas and insects from living in your carpets, sprinkle with any common household salt. Leave for about ten minutes, then hoover it up. This is a far safer way than using chemical flea powders.

Flea Shampoos

An effective flea shampoo can be made from a cheap supermarket brand of shampoo mixed with a few drops of pennyroyal or eucalyptus oil, which can be bought from a chemist or aromatherapy shop. The oils will last you well over a year.

FLIES

We do not want to use commercial fly sprays when we have animals around the home, as it could be dangerous if our pets eat the sprayed flies. We can use the following quite safely in our homes and kennels: tie a wineglass, broken at the base of the stem, to a light pole, long enough to reach the ceiling. After sunset or in the morning when the flies are asleep on the ceiling, hold the glass, half full of methylated spirits, under the flies. The fumes will cause the flies to fall into the spirits, killing them instantly. The spirits can be poured off into a bottle and used repeatedly. A room can be cleared of flies this way in two minutes.

Another tip is to put some eucalyptus oil onto a cloth, open the door wide, and wave the cloth vigorously, working towards the door. The flies will disappear rapidly.

To prevent the entrance of flies into a room hang up a fresh bunch of stinging nettles in the window.

A bunch of walnut leaves hung up in a room also discourages flies.

Cedar wood oil (obtainable from any chemist) that is placed on cotton wool and left about the rooms will keep flies away.

Hang fresh bunches of elderleaves in your room in

summer to banish flies.

Soak some small pieces of sponge in warm water, then sprinkle with a few drops of oil of lavender. Place them about the room out of sight, or behind pictures. The smell, which is not unpleasant, will keep flies away.

GRASS CUTTINGS

Dogs are always bringing cuttings indoors on their coats. If your dog rolls and leaves a grass stain on your carpet, brush the affected area, then using a weak solution of methylated spirits and cold water, start rubbing on the outside of the stain and work your way inwards towards the centre. Allow to dry and then brush the nap.

GRASS STAINS ON DOG'S BEDDING

Just rub plenty of treacle into the stained area and leave for about ten minutes, then wash in the usual way in tepid water.

Another good way of removing grass stains from your dog's bedding is rhubarb juice: cut a stick of rhubarb into chunks and place them into a saucepan of cold water. Boil briskly for a few minutes and then hold the stained bedding in the hot mixture for a few minutes. Rinse off with cold water. Then wash out the bedding in the usual way.

GRASS SEEDS

Please check in-between the toes of your dogs in summer for grass seeds in the paws. It is remarkable what these seeds can do. They work their way into the gap between the toes, like needles, and sometimes they disappear before you have time to notice them. Keep all hair short under the pad and in-between the toes. Then check by spreading the toes of your dog and feeling with your fingers for any grass seeds. If you can still see one, first rub in some vegetable oil to loosen the skin, and then remove the seed with tweezers. Apply some honey over the wound to encourage quick healing. If the wound becomes infected, you must go to your vet.

If burrs are caught into the coat, again rub vegetable oil into the coat and very gently comb through.

GROOMING

Burnt Coats

If your dog lies too near the fire and singes it's coat, you get this horrible smelly burnt patch on them that they never seem to notice! Just get some mayonnaise from the fridge, rub it into the affected area, and leave for ten minutes to soak into the coat, then rinse out with tepid water. If the water is too hot, you will end up with a mess. Allow to dry naturally, and brush through.

If the dog's skin is slightly burnt, clip away the hair and apply strong lukewarm tea on a compress. The tannic acid content in the tea helps aid healing.

Conditioners for Coats

Some dogs, no matter what we do, never seem to be able to get a glossy coat. We can use several cheap, natural things without spending a fortune. A healthy food addition is to save water from cooked vegetables to add extra flavour and vitamins to your dog's dry food. Allow the water to cool down before adding it to the food.

A good conditioner is to add a raw egg or two tablespoonfuls of melted animal fat to a dog's daily diet, if the dog is of Retriever size. A dessertspoonful is enough for smaller dogs. These additions will help produce a shiny coat. Vegetable oils do not have the same effect. Also, next time you open a tin of tuna for yourself, save the oil and pour it over your dog's food for health and extra taste.

To make a wonderful coat conditioner, mash together an old banana with a rotten avocado and rub this into the coat. Leave for about ten minutes then rinse out. You could also use mayonnaise rubbed in and left for the same time before rinsing out. Mayonnaise is an excellent conditioner, which I have used many times on my show dogs.

Just before the last rinse, pour any leftover beer or a dilution of vinegar over the dog to give a shine.

Dirty Coats

Dogs' coats can become very dirty and we don't want to bathe them every time we bring them home. A handy, harmless way of keeping coats clean in-between baths is to rub bicarbonate of soda into the coat, and then brush it out. The bicarbonate of soda acts as a deodorant and as a dry cleaner.

Dry Coats

Most dogs love to lie in front of the fire or rayburn. This can make their coats dry and dull and especially in cold wet weather. Treat the coat with Lemon Tonic (p.71). Another good coat conditioner is to put one an a half tablespoons of instant dried milk powder to one gallon of warm water.

When you do bathe your dog, place an old tea strainer over the plug. This will trap all the dog hair and stop the plug from being bunged up.

Oil or Tar on Coats

Dogs can pick up oil on their coats from various sources: walks on the beach, oil patches on the drive from the owner's car, etc. Eucalyptus aromatherapy oil (available from health food shops and chemists) on cotton wool is excellent for removing this.

GROOMING BRUSHES

Your dog's grooming brushes and combs must be cleaned regularly to keep them pest-free and clean. Mix 2 cups of hot soapy water with half cup of vinegar. Soak them for half an hour in this solution then rinse clean. This will sterilise and clean your grooming brushes.

If you have a very fidgety dog that just will not stand still for you to brush him, you could try this. Smear some marmite on a smooth, easily cleaned surface like the front of a door, stand your dog in front of this and he should be so busy licking this off that you should be able to get a few minutes of brushing done! Always disinfect the surface after wards.

HAIR ON FURNITURE AND CARPETS

Take a trainer (preferably a non-smelly one) put your hand inside, rub the sole over the carpet, and then vacuum it afterwards. This will remove all the stubborn hair.

For furniture (and clothes) use sellotape wrapped

round your hand with the sticky side out and rub this over the furniture and clothes - a lot cheaper than buying special brushes.

If you want your carpet to smell nice, don't buy expensive carpet deodorisers as some contain chemicals that can cause serious skin problems to children and pets. Instead, sprinkle a cheap talcum powder over the carpet, then hoover it up. This does the same thing for a fraction of the cost.

HAYFEVER AND DUST COUGHS

Our dogs can get a form of hayfever or dust allergy coughs. For this, you can make up your own cough medicine. Take a cupful of diced swede or diced onion and soak it in a cupful of brown sugar overnight. Keep this mixture in the fridge and feed 1 tablespoonful of the liquid to your dog each day until the symptoms clear.

Dried sage and honey mixed and added to the food once a day is also effective.

HEAT PADS

I have noticed that you can now buy microwave heat bags for your dogs. You pop them in the microwave for a few minutes then you put them under your dog's bedding. It keeps the bed warm all night.

This is a good idea, but all the ones I have seen are very expensive.

Why not make your own home-made ones for free! You have to act on this hint from early summer, because it will probably take you all summer to collect enough fruit stones to make one. You will need to keep and dry small fruit stones: cherry stones are best. When you have collected about 1lb in weight, sew them into a cotton bag. This can be heated in the microwave for a few minutes, and the heat from the cherry stones will then last all night, exactly the same as an expensively bought heat pad.

Another very good way of keeping a dog's bed warm overnight is to cut some old "bubble wrap" into the shape of your dog's bed and place it underneath the bedding. This makes a lovely cheap and efficient thermal blanket.

HICCOUGHS

Your dog may sometimes develop hiccoughs. First make sure that worms do not cause them. If not, it may be that your dog is greedy eater and bolts down its food. Just give your dog half a teaspoon of baby's gripe water.

Alternatively, try giving 1/2 teaspoon of sugar or half a teaspoon of vinegar by mouth.

HOT WEATHER

Keeping Dogs Cool

If you have a dog (especially if it is black or old) it can get distressed in hot weather. Try soaking an old towel in cold water the night before, then fold the towel up, retaining as much moisture as you can, place the towel in a plastic bag and put it in the freezer. The next day you start off by placing the it under the dogs bedding, then gradually as it begin to thaw out take the towel out and place it over the body of your dog. This way it should last about four hours.

If you don't have a freezer, only a freezing compartment in your fridge, make as many ice cubes as you can. Then 3/4 fill a hot water bottle with cold water; top it up with the ice cubes and use in the same way as a towel.

Never take your dog out in the car when it is hot, if it can be avoided. You may park in the shade but as the sun moves round (even with the windows open) the poor dog can literally cook (like being roasted alive). It does not take very much sun to make the temperature inside a car reach dangerous levels. If you left a tomato on your dashboard for a couple of hours, by the time you returned the tomato would have burst open. So think what effect heat can have on your dog if you left him... even with the windows slightly open, if it is 72 degrees outside, within half an hour the inside temperature can be over 100 degrees. As dogs cannot sweat like us, so they will die of heat exhaustion very quickly.

If the dog has to travel with you, a hatchback or an estate car can be turned into quite a cool travelling area.

The night before you are travelling, fill large empty plastic lemonade bottles with water and place in the freezer overnight. Wrap the frozen bottles up in old towels and place around the inside edge of your car's tailgate. This will help to keep your dog cool for a long time while travelling.

A remedy for heat stroke is one third of a teaspoon of bicarbonate of soda in the food bowl of large dogs. This will help in hot weather.

Dissolving 1 teaspoon of salt and 1 tablespoon of sugar in 1 pint of boiled water can make a good re-hydration drink. Place in an airtight jar and give small amounts frequently.

If your pets are going off their food in hot weather and you are worried about them getting the proper vitamins, make some 'ice lollies'. Dissolve a stock cube in a pint of water then freeze in the ice cube container and put a few of these at a time in a bowl. Most cats and dogs love them.

ID TAGS FOR HOLIDAYS

Many of us will be taking our dogs on holiday with us. They should always be wearing their normal metal identification disc with their home phone number on it.

Here is a little tip for whilst you are away. Buy a cheap plastic tube and put your holiday contact address inside, then attach this to your dog's collar. If your dog gets lost on holiday, valuable time can be saved getting him back to you.

INSECT BITES

Insect bites can really annoy your dog. A quick cure is to get a piece of soap, dip it into cold water and rub the surface of the bite. The irritation will stop immediately.

JAUNDICE

This is a bile disorder shown as a yellowness of the skin and bodily fluids.

Dandelions made up in the same way as 'chickweed tea' helps with jaundice and cleanses the body of toxins. Just put a couple of drops of the tea in your dog's food at every meal, and the jaundice will improve after a few days.

KENNEL CARE

A quarter of a pint of lavender water diluted in a gallon of water can be used as a wash on kennel floors to keep the flies away.

Salt sprinkled onto concrete runs is more effective than disinfectant for killing flies' eggs.

After washing kennel floors, add a cupful of vinegar to a bucketful of water for a good antiseptic rinse.

Keep a bowl of vinegar in the kennels to keep "doggy" smells at bay.

Ants

To prevent ants coming into your kennels, rub the juice of an orange around the doorways. The ants won't venture over this. You could also try sprinkling dry mint or chilli powder across their path.

Another tip to prevent ants coming into your kennels is to put fresh onion skins by the doorways.

Cleaning

Once a week, instead of just washing my dogs' feeding bowls and chopping boards I like to disinfect and deodorise them. Forget the expensive shop-bought products: just rub the item that you are cleaning with ordinary bicarbonate of soda, then spray on full-strength vinegar. It will bubble and froth as these two natural chemicals react. Let it sit for 5 minutes, then rinse off in clear water.

To remove the smell of dog food from feeding bowls, add a spoonful of dry mustard to the washing-up water.

Mice

If you are troubled with mice in your kennels, sprinkle oil of peppermint around. Mice hate the smell and will not come near it.

Slugs

To deter slugs from coming into your kennels, up-turn empty grapefruit shells outside the doors and slugs will go into them and not your kennels.

KENNEL COUGHS AND COLDS

Kennel cough sounds dry, harsh and frequent. It can be serious in puppies and older dogs and the owner must seek a vet's advice. In mild cases the following recipe can help enormously: mix one tablespoon of Friars Balsam with 1 pint of boiling water in an old ice cream container or similar. Cover with a secure perforated lid and place in the dog's cage with the dog, draping a towel over the cage. Do this twice a day for 5 minutes at a time. The vapour will ease your dog's cough.

KIDNEY PROBLEMS

Increased thirst and urination can suggest kidney problems and, again, should be diagnosed by a vet. Dogs are more prone to kidney problems as they get older. To try and delay this problem, try a pinch of fresh or dried ginger in the diet every day, or you can finely chop some water melon peel and add this daily to your dogs diet. You can dry watermelon peel naturally, then liquidize and store it in an airtight jar for use later.

LEADS

Personally, I much prefer well-made nylon collars and leads, as they do not discolour the coat when wet.

Getting a puppy used to wearing a collar and lead is one of the earliest training tasks you will have. One gentle way to manage this is as follows: Put the collar and lead on the puppy just before its feed time. Place the food on the far side of the room where the puppy can see it, hold the lead while the puppy goes over to the food and let it trail on the floor while the puppy is eating. Repeat this over several days and the puppy will soon learn that the collar and lead mean something pleasant.

Another handy vitamin pack is made up as follows:

✳✳✳✳✳✳✳✳✳✳✳✳✳✳✳✳✳✳

1 teaspoonful salt, 1 dessertspoonful of glucose (from chemists or cake decorating shops), and 1 tablespoonful of cheap margarine. Mix this together, then freeze into ice cube trays. Place one cube in your dog's drinking water.

LONG EARS

Long-eared dogs are always getting food in their ears when they are eating. There are two ways of dealing with this. The first is to find a plant pot, into the top of which you can securely fit the dog's feeding bowl. Plug up the hole in the bottom of the pot and half fill the plant pot with soil or something to make it stable. Place the bowl on top. When the dog eats, his ears will fall to the side of the bowl and stay clean - you can do this with the water bowl as well.

If your dog is a quick eater, another method is to cut a section from leg of a pair of ladies' tights - about 10" to 12" long - and slip this over the head (like a snood) so that the ears are laid back against the head. Your dog can then eat without getting messy ears and look like Batman into the bargain.

MANGE

This is caused by a parasite, which can also infect

✳✳✳✳✳✳✳✳✳✳✳✳✳✳✳✳✳✳

humans. Keep an eye on your dogs as many foxes are being affected with a particularly nasty form of mange, which can be passed onto domestic animals if they are exposed to areas where foxes rest. Demodectic mange is caused by a mite and has been killing foxes in the South West of England for some time. Dog owners should be wary if foxes visit their gardens. Foxes rarely survive mange, but domestic animals can recover under veterinary care. If your dog starts to show sores between the toes, over the eyes or on the nose and bald patches begin to develop, get in touch with your vet.

MEAT

To keep Fresh

If you only use half a tin of dog food or if you cook your own dog food, sprinkle a few coffee granules on top of the food and it will stay fresher for longer.

To Tenderise

Cooked meat is preferable to raw for dogs because the cooking kills bacteria and worm eggs. It should, however, be mixed with some sort of roughage such as cooked rice (brown, preferably), pasta or dog meal. Some people still cook meat for their dogs themselves, but the meat sold for pet consumption is usually taken from the toughest cuts. To tenderise the meat, add a pinch of bicarbonate of soda to the cooking water and simmer gently.

NAILS

Bleeding Nails

If your dog breaks a nail and it is bleeding rather badly, which it will do if the "quick" (or vein) has been broken, and you have nothing to hand to stop the bleeding, rub the nail over a bar of damp soap. This will arrest the bleeding.

Distilled witch hazel, easily available from chemists, is one of the best coagulants that I know of. It will stop bleeding on cut pads, and is safe to use on most dogs. It is also very good on minor sprains. In both cases, just liberally dab onto the affected area, and allow to soak in.

White pepper shaken onto a bleeding claw will stop the flow. Otherwise, ordinary flour will work as well. Just dip the toe into the flour and pat it into the claw.

Weak Nails

If your dog's nails split and crack when they get to a certain length, dab surgical spirit or cider vinegar onto the nails daily.

✳✳✳✳✳✳✳✳✳✳✳✳✳✳✳✳✳✳

If your dog suffers from soft nails, try soaking the nails up to three times a day in iodine for as long as you think fit. They should very soon get hard.

You should also include more calcium in your dog's diet.

NETTLE STINGS

Bathe the stings with cold tea on cotton wool.

A handful of fresh parsley or dock leaves (the long, broad, rusty green leaves usually found growing near nettles) rubbed well into the site until the sap flows will take away the irritation immediately.

NIGHT-TIME SAFETY

Reflective Strips

You can now buy reflective collars and leads for walking your dogs at night, so the cars' lights can pick you out, but you do not have to pay a lot of money for this.

Good car accessory shops and cycle shops sell rolls of reflective tape very cheaply. If you stick a strip of this on your leads and collars, it will work just as well. You can even wrap a small bit around your cat's collar or put it on the back of your own coats and gloves.

✳✳✳✳✳✳✳✳✳✳✳✳✳✳✳✳✳✳

NOSES

Cracked

A common problem is cracked noses on some dogs,and it is quite an unsightly problem. There are a couple of things that you can use for this: ordinary cheap glycerine smeared on the nose in the daytime and a lip salve at night. You should see a great improvement within 10 days.

You could also try bathing the nose twice a day with one teaspoon of T.C.P. in half a teacup of boiled cooled water.

In summer you can try using a sun block: zinc oxide cream can be beneficial - available from a chemist.

Cows' udder cream (usually available from farmers' suppliers and country chemists) has also been known to help.

Loss of Pigmentation

If your dog loses the pigment on its nose (or from its pads or eye rims) it is nothing to worry about, as this just a cosmetic problem. However, it does concern some dog owners. This colour variation can vary from season to season but in some cases the loss of nasal pigment has been attributed to the use of plastic feeding and water bowls. So if you are worried, change the plastic bowls for pottery ones and see if this makes any difference. If you are lucky enough to obtain elderberries, dry them out thoroughly and slowly, then pound them to a powder. Store this in an airtight jar and add half teaspoonful to food daily to aid pigmentation. Using the same method with seaweed can also help.

NURSING BITCH

If your bitch has had puppies, help her to produce more milk by warming up (do not boil) crushed oats with water. Mix up as much as you think that she will eat. If she will not eat the mixture on its own, mix it in the normal feeds in small portions at a time.

PILLS

There are various expensive gadgets on the market to enable you to administer tablets to your dogs, but you do not need them. Place the pill in a good knob of butter or margarine, place the butter in the side of your dog's mouth right at the back where there are no teeth. Your dog will not be able to bring the pill back up, as it is too slippery.

POISONS

Household chemicals, pesticides and medications all

present risks of poisoning to dogs. Signs of poisoning are burns in and around the mouth, vomiting, profuse salivation and convulsive seizures.

Poisons fall in to five categories: corrosive, irritant, narcotic, hypnotic and sedative. Steps to take in case of suspected poisoning are:

1. Phone your vet immediately for advice as to whether or not to induce vomiting

2. The most effective way to induce vomiting in dogs is to administer sodium peroxide, available from chemists (1 - 3 teaspoons every 10 minutes, no more then three times) until vomiting occurs.

3. Do not induce vomiting if your dog swallows an acid, alkali, solvent, or a heavy-duty cleaner; is severely depressed or comatose; swallows a petroleum product; swallows tranquillisers; swallows sharp objects, or if more than two hours have passed since the poison was swallowed.

4. If your vet is unobtainable and you know what your dog has ingested, your local poison control hotline may be able to advise you. In any case, get your dog to a vet as quickly as possible, taking along a sample of the substance it has eaten, if this known,

preferably in its original container. The container may help your vet determine appropriate treatment because it should provide information on chemical composition, and possibly give instructions on what to do in cases of poisoning.

5.   If you see blisters around the dog's mouth, it has probably ingested a corrosive substance. Acid corrosives are found in car batteries and etching solutions. Flooding the burned areas with water, and then administering a dilute alkaline solution of bicarbonate of soda and water should treat acid poisoning.

6.   Alkaline corrosives are found in ammonia and in sodium hydroxides used in preparations used to unclog drains. After flooding the area with water, treat with a dilute solution of water mixed half-and-half with lemon juice or vinegar.

7.   After administering a dilute solution to combat acid or alkaline poisoning, try to get your dog to drink something. Milk, raw egg white, or olive oil are effective demulcents that will soothe the inflamed tissues. Take your dog to your vet for further treatment.

POISONOUS BERRIES

In autumn, when there are a lot of berries around and if you suspect that your dog has swallowed something poisonous, the old remedy is to make it drink some salty water. But as this does not taste very nice it is not always easy to do.

If you place a good pinch of salt on your dog's tongue then give some fresh water, the offending berry should appear. If you think the berry is poisonous, pop along to your vet for a check-up. Remember to take the berry with you.

At Christmas we have many plants in our homes that we don't usually have: mistletoe, holly, poinsettias, Christmas cherry, etc. If you have young dogs around make sure you pick these berries up when they drop because they are very poisonous.

PREGNANT BITCH

When a pregnant bitch feel a little sickly, try feeding her a ginger biscuit to balance her tummy. Raspberry leaf tablets are also said to ensure an easier birth.

PUPPY CARE

When you bring a puppy home and he misses his

mum, sprinkle his bed with sweet marjoram essential oil. Put 2 drops on your hand and put on the puppy, avoiding the eyes. This will calm him down.

At night, wrap an alarm clock and a warm hot water bottle in an old jumper, and put this in the puppy's basket. The ticking will sound like the mother's heartbeat and will soothe the puppy.

Sore Bottoms

Always keep an eye on young puppies, as their little bottoms must be kept clean at all times, to reduce any risk of infection. When puppies are getting used to new diets as they get older, they also sometimes get sore bottoms. Smear a little vaseline around the anus: this will quickly solve the problem.

RINGWORM

This fungal infection is quite rare in dogs. Tell-tale signs are patches of hair loss, with a greyish scaling and crusting at the edge of the patch. Sharing infected collars or grooming equipment generally causes it. Apply cider vinegar on ringworm patches up to six times a day until the condition clears.

RUMBLY TUMMIES

A child's dose of Milk of Magnesia will usually stop

the noises within an hour.

Baby's gripe water, given one teaspoon at a time can also help, as can a dessertspoon of oatmeal added to food.

A drink of goat's milk with a spoonful of honey added, given night and morning, may work. You could also try a teaspoon of natural yoghurt night and morning, and starch meal of pasta, brown soda bread, or rice with a sprinkling of cheese last thing at night.

SKIN

Preventing Blemishes

To prevent blemishes and spots on short-haired and hairless dogs, mash 3 strawberries into a quarter of a cup of vinegar, let it stand for at least 2 hours, then strain the vinegar through a cloth. Pat the mixture all over the dog and leave as long as you can then wash it off. The skin will soon be free of pimples and blemishes.

Rough Skin

First apply glycerine on cotton wool, then apply ordinary household salt. Leave for 1 minute then rinse off with tepid water and dry.

To remove rough skin kindly instead of using a

harsh abrasive, rub over with burnt toast.

Skin Tonic

In winter with the cold drying winds and wet weather, dogs' coats can get out of condition, especially when we let them dry in front of the fire. The next time your dog needs a bath, try the following for a nice skin tonic: Take one whole lemon, and one pint of nearly boiling water. Thinly slice the lemon, including the peel, and add to the water. Let it steep overnight. The next day, sponge this into your dog's coat and let it dry. This tonic is good for skin problems and fleabites.

SMELLS

On Carpets

If you have had a puddle from your puppy or kitten on your carpet and have cleaned it up but can still smell that unmistakable smell, put a bowl of neat vinegar in the room overnight and the smell will have completely disappeared by morning.

If your puppy has an accident on your carpet and it leaves a light carpet stain, dissolve 2 tablespoons of salt in half a cup of vinegar. Rub this into the carpet stain and let it dry. Then vacuum up the residue. Your carpet will smell fresh and the vinegar will not lift the colours.

You should never use pine disinfectant with dogs as the pine scent compounds badly with the urine and makes the smell worse. Use a product containing chlorine instead, or use diluted vinegar as a disinfectant.

You can make your own, environmentally friendly air freshener that won't harm your dogs. Put the following in a plant spray bottle: 1 teaspoon of bicarbonate of soda, 1 tablespoon of vinegar, 2 cups of water. After the foaming stops, put on the lid and shake well. Use this as a very effective air freshener where there are dogs around. You could also add a teaspoon of cinnamon to this recipe, if you like the aroma.

Fox

Sometimes when we take our dogs out for a walk and let them off the lead, they find a fox urine smell on the grass and absolutely love to roll in it - leaving them stinking of fox! The quickest way of removing this smell is to put some tomato ketchup on the affected area, leave for a minute or two then rinse it out. This will not stain the coat and the smell is lifted out instantly. Alternatively, pour a bottle of vanilla extract into a pint of water. Pour this over your dog and leave it to soak in for about ten minutes, then bathe in the usual way. All smells will then have disappeared.

Smells on Hands

If your hands get very smelly after preparing fresh food for your dog, just rub a little mustard powder into your hands, then rinse. All the smell of the pet food will have been taken away.

Smells in the Microwave

If we are cooking food for our dogs in our micro-wave ovens, fish or tripe for example, we want to get rid of the smell as soon as possible. To do this, mix quarter of a cup of vinegar and 1 cup of water in a small bowl, and heat for 5 minutes. This will remove odours and soften baked-on food splashes in your microwave.

Smells in the Refrigerator

Fresh meat and fish can leave your fridge very smelly. A very good way of deodorising the fridge is to leave a bowl filled with fresh cat litter in the fridge. This will absorb all smells. Change the litter in the bowl weekly.

SNAKE BITES

Fortunately these are very rare, but they can happen on commons and in the country. If you think your dog has been bitten, look very carefully at the swelling, for two puncture wounds. Take no chances and always carry a small pot of Permanganate of Potash. It can be obtained from any chemist at very little cost. It is the best antidote that I know of. Just sprinkle a few crystals into the wound. As the potash starts working straight away it gives you valuable time to get your dog to the vet.

Also try to buy the houseplant Aloe-Vera. This plant is invaluable: all you need to do is break off a leaf and use the juice on any wound. You can keep the leaf fresh in a plastic bag for some time in the fridge and cut it and use it again.

SORE SPOTS

The humble lemon that most of us have in our kitchen can be very useful on our dogs. For slowly healing sores, dab on some fresh lemon juice. This will greatly aid healing. Neat lemon juice dabbed onto fleabites or stings will do the same thing.

Aspro - not aspirin - crushed and dusted onto runny or weepy spots and sores will quickly help them heal.

SUNBURN

Pour half a cupful of hot milk onto a thick slice of lemon, leave for one hour, them strain the mixture, throw away the curd and use as a lotion the clear liquid that remains.

Half a pint of lukewarm water to which a cupful of cider vinegar has been added can also be dabbed onto sunburn for relief.

TEETH AND MOUTH CARE

Tartar building up on teeth will soon cause tooth decay. To stop this, we should try to clean their teeth regularly. A lot of dogs do not like a tooth-brush in their mouths, so try wrapping a fabric plaster around your index finger, wet it and dip it in a little bicarbonate of soda and gently rub around your dogs teeth.

Another good tooth cleaner is mashed strawberries. Believe it or not, strawberries clean teeth very well. That goes for humans too!

Clean dogs' teeth naturally by giving them a large raw carrot to chew on.

If your dog has toothache, of course you have to see your vet, but in the meantime a teaspoonful of lemon juice should relieve the pain.

Chop and cook in a small quantity of water the parts of fennel you do not use. Liquidize the entire contents, including the water, cool and store in an airtight jar in the fridge. This will keep for about

two months. Rub a cloth soaked in this mixture around the dog's teeth and gums to clean them of bacteria, and keep the breath smelling sweet.

Mouth Ulcers

If your dog develops small, whitish sore spots on its gums, these will be mouth ulcers. They often occur when puppies are teething. If you apply a wet tea bag to the ulcer once or twice a day, this will help it heal more quickly, because black tea contains tannin, which is an astringent. If the new teeth are driving your dog mad, gently smooth some baby teething gel onto the gums to help with pain relief.

A little nutmeg rubbed into the affected spot will also help greatly with the pain.

THIN DOGS

If you have a dog that won't put on weight and you have had it checked by a vet to ensure that there are no medical reasons for this, there is a sure-fire way of putting the weight on. Soak those quick cooking oats in milk for at least 12 hours or overnight, then add more milk in the morning with a pinch of salt and feed raw once a day along with the normal diet until your dog reaches it's correct weight.

Dogs need unsaturated fatty acids in their diet to maintain good health. The easiest and cheapest way of providing this is to put a teaspoonful of ordinary cooking oil in the food.

Make sure your dog sticks to its normal diet over Christmas, as a sudden change of diet or too many rich tit-bits could result in a very upset tummy. Also during the Christmas period there is always lots of chocolate about. Don't under any circumstances give a dog plain chocolate, as this contains theobromine, which irritates a dog's stomach and increases it's heart rate. Avoid this, as the dog can become very ill. Milk chocolate does not contain this chemical.

The following dietary tonics are worth trying:

young nettles are regarded as an unwanted weed in our garden, but they are rich in iron and vitamins, and so can provide an excellent vitamin supplement for dogs without spending any money at all. Carefully pick out the young fresh tops from nettles, then blanch them quickly in boiling water and add to your dog's food. Alternatively you can dry them as you would herbs, crumble them up and store them in a jar. Add a pinch at a time to your dog's food.

Baby dandelion leaves treated in exactly the same way as the nettles are equally as good.

Folic acid is to be found in green vegetables, so

these home made tonics are particularly good for a pregnant dog (and, of course, pregnant women).

Hungarian red pepper (i.e. paprika, not chilli pepper) sprinkled sparingly onto an old dog's food is a very good vitamin powder. Surprisingly, dogs do like it!

TICKS

Ticks are very tiny when they first attach themselves to the dog. You will only spot them as they fill with blood.

If you dog gets a tick on him you will soon know about it. The tick attaches itself and as it drinks the blood it swells up, looking like a pea, but blue-red in colour. The way to remove ticks is to put surgical spirit on them to make them lose their grip (you can use whisky if you want them to die happy). Wait about ten minutes, and them pull and twist as near to the skin as possible, making sure that you remove the tiny head. If this is left in the skin it can set up a severe infection.

You could also suffocate it by painting some nail varnish on it. After a few minutes the tick will drop off.

I would not recommend trying to remove a tick without putting anything on it first.

Once you have removed the tick, bathe the area with a little vinegar water.

TOYS: PLAY BALL

There is new, very good toy on the market: an "Activity Ball" designed with a couple of holes into which you put a few doggy treats. You give this to your dog to play with to keep it amused whilst you are out for a couple of hours. These cost about £10, but you can make your own.

Take a tennis ball, cut a slit or make a small hole in it and push a soft chew or small dog biscuits inside and your dog will play for ages with this.

We often give our dogs empty plastic containers to play with. This is fine, but never give your dog an empty dishwasher liquid container, as dishwasher liquid is poisonous to animals.

UPSET STOMACH

With a sudden change of weather dogs are prone to minor tummy upsets. There are a couple of things that you can use for your dogs quite safely. Get the cooked white of an egg and cut it up with a knife and a fork. Mix with one tablespoonful of water, and then feed to the dog a small spoonful at a time. If you have a very small dog or a dog that will not take this mixture, try giving very flat soda water to drink. Both of these treatments should effect a change within 24 hours.

Homemade pet lucozade:

2 dessertspoons of glucose, (from a chemist) and half a teaspoon of salt, made up to a pint with boiled water. Cool and administer in hot weather.

If your dog has a slight upset tummy and diarrhoea, a very quick cure is to mix some live yoghurt (it must be live) with honey, in the following proportions: 1 tablespoon of yoghurt & 1 teaspoon of honey for medium-sized dogs (adjust the quantities accordingly). The problem should clear up within 24 hours, and the mixture is also safe for puppies. Of course if the diarrhoea persists, you must take your dog to the vet.

A teaspoon of fresh chopped coriander mixed in with food will settle a tummy.

Flat soda water is also good for tummy upsets.

Try the following mixture:

One teaspoon of brandy
One teaspoon of sugar
One tablespoon of water

Mix thoroughly and give one teaspoon every fifteen minutes until all the mixture is used up.

VITAMIN POWDER

There is no need to buy expensive vitamin powder for your dog as you can make your own very cheaply, and it is just as good. You will need the following:

Two cups of dried yeast
One and a half cups of bonemeal
Half a cup of kelp powder

Mix the ingredients together and store in a sealed jar in the dark in a cool place. Just use a very small pinch at a time, as this is all that would be needed. All good chemists, garden centres (for the bonemeal) and health food shops will sell the ingredients very cheaply.

WARTS

As dogs get older they can contract warts. Although harmless, they are annoying as when we are grooming our dogs we can inadvertently make them bleed. There are two natural cures that work very successfully, one will cost you a small amount of money and one is completely free.

You could squeeze the contents of a vitamin E capsule onto the wart daily. However, the free way to deal with warts is to pick a fresh dandelion and dab the wart with the white juice that oozes from the stem. Do this daily until the wart disappears.

Raw garlic or banana skin can also get rid of warts.

Grind peanuts with brown sugar, pour on boiling water, and let the mixture cool. Administer three teaspoonfuls daily.

WASPS AND BEES

If your dog likes to sunbathe out in the garden all day, prevention is better than cure. Spray some red wine vinegar lightly over the dog to keep the bees and wasps at bay.

The best way of preventing wasps from harming your dogs (and ourselves) is to make home made wasp traps. All you need is an empty jam jar with a lid. Cut a cross shape in the lid and bend the flaps inwards. Smear jam or any sweet substance in the jar and fill the jar to about 1 third full with water. The wasps fly into the jar to get at the jam but cannot get out again. You can use any left over beer and this way they die happier!

A wasp will sting and then fly away, but a bee sting will be left behind. This sting must be removed as the poison sac is attached to the sting and continues to pump poison into the wound. To treat wasp or bee stings on the body of your dog, use vinegar and dab it on with cotton wool. If you have any whisky you can use this, as it will give instant relief. A cold tea bag or a teaspoonful of bicarbonate of soda mixed with a little drop of water and applied on the sting will relieve it greatly. This is also good for nettle stings and minor burns as well. Try rubbing damp soap on for instant relief.

If your pet gets stung anywhere inside the mouth, immediately give an antihistamine tablet (half a tablet if it is a small dog or cat) and then take it straight to the vet.

WEED KILLERS

Care should be taken to avoid harmful weedkillers in the garden, as they may be harmful to pets. Many of us have got patios or concrete areas where our pets like to sunbathe, but the grass and weeds will grow through the cracks, and if we pull them up they will still grow again. A method that I have used successfully over the years is to pour undiluted white vinegar into the cracks. This will not harm your dogs, and will kill the weeds safely for you.

WET BEDDING

Dogs' bedding can become damp and smelly because of walks in the rain, swimming, incontinence etc. This makes the bedding an ideal breeding place for germs. When you are washing the bedding, add along with the usual soap a quarter cup of vinegar. This will inhibit mould and fungus growth and kill germs.

If you sprinkle bicarbonate of soda liberally into the bedding then rub it in and leave it overnight before washing, all traces of the smell will have gone after the wash.

WET DOGS

The quickest way of drying dogs is with damp chamois leather. Rub the dog all over, rinse the cloth out, and keep repeating the process until the dog is dry.

Lard or washing soda rubbed into the coat will also remove oil or tar.

WHITE DOGS

These dogs get discoloration around the eyes. There are some very expensive eye lotions on the market for this, but they aren't necessary. All you need is a bottle of witch hazel that you can get from the chemist for about 65p. Dilute 1 part witch-hazel to 2 parts water and wipe this under the eyes, using a pad of cotton wool. Do this every day and gradually the stains will disappear.

If you don't want to buy witch hazel there is an alternative way that is even cheaper: use bicarbonate of soda. Mix 1 tablespoonful of bicarbonate of soda with 1 pint of warm water. Keep this in a sealed jar. Dip cotton wool into this daily and wipe around the eyes and other stained areas. The staining will visibly disappear.

Mix fullers earth (white if you can get it, from a good chemist) and boracic powder (also from a chemist) with milk of magnesia to make a thick paste, and apply to the affected area.

An alternative is a dry mix of boracic powder and cornflour, rubbed into stain daily.

WORMS IN DOGS

You will not see worms in dogs' droppings until they have been wormed, because until the worming pills have taken effect, the worms will still be living in your dog's intestine. All adult dogs must be wormed every six months, without fail. I also give my dogs one garlic tablet every day, and I can honestly say that I have never had a worm problem with them.

You can buy garlic tablets in chemists, but you can make your own preparations. Take 1oz of fresh crushed garlic to 20 ml of boiled water, put in a glass jar after cooling and seal. Add 2 drops to your dog's food each day. Made this way it will keep in the fridge for about two months, or alternatively you can freeze it into ice cubes and use as required.

2 pumpkin seeds for 4-5 days every three months helps worms release their hold on the intestine.

Useful Addresses and Further Reading

British Association of Homoeopathic Veterinary Surgeons c/o Alternative Veterinary Medical Centre, Bhinhan House Stratford in the Vale Faringdon, Oxon SN7 8NQ tel: 01367 718115

Association of Chartered Physiotherapists in Animal Therapy (ACPAT) Morland House, Salters Lane nr Winchester, Hants.tel: 10962 863801

Epilepsy in Dogs: The Phyllis Croft Foundation 3 Spring Close, Gt. Harwood, Bucks MK17 0QU

Pet Loss Befriendment Service tel: 0891 615285 (internal calls to this helpline cost 50p per minute). After 31.12.1998 the number will change to: 01877 330996 **(The Society for Companion Animal Studies)** 10B Leny Rd Callander FK17 8BA

Natural Health for Dogs and Cats Richard Pitcairn D.V.M. PhD, and Susan Hubble Pitcairn. Prion Press isbn 1-85375-091-3   1982

The Doctor's Book of Home Remedies for Dogs and Cats By the editors of **Prevention Magazine Health Books** pub. Rochdale Press, Inc. Emmaus, Pennsylvania isbn: 0-87596-294 1996

✳✳✳✳✳✳✳✳✳✳✳✳✳✳✳✳✳✳

Other titles by Broadcast Books

Ain't Misbehavin': A Good behaviour Guide for Family Dogs by David Appleby

£9.95pbk illustr. 288 pages isbn: 1 874092 7 29

Using proven, kind and simple strategies, David Appleby will help you turn your unmanageable dog into the perfect family pet. Highly acclaimed by experts.

"*Excellent reading, full of knowledge and helpful advice*"
Dog Training Monthly

"*I was completely converted. It is utterly commonsensical*"
Sunday Telegraph

Understanding the Border Collie by Carol Price

£9.95 pbk 276 pages illustr. isbn: 1 874092 86 9

The first book of its kind for all who want to learn more about this old and noble breed and to become successful owners of happy dogs. Advice on how to choose your dog - whether rescue collie or collie cross, or pup - and how to rear, feed and train it.

(published in February 1999)

✳✳✳✳✳✳✳✳✳✳✳✳✳✳✳✳✳✳